John Harvey Kellogg

The Influence of Dress in Producing the Physical Decadence of American Women

Annual Address Upon Obstetrics and Gynecology

John Harvey Kellogg

The Influence of Dress in Producing the Physical Decadence of American Women
Annual Address Upon Obstetrics and Gynecology

ISBN/EAN: 9783337379490

Printed in Europe, USA, Canada, Australia, Japan

Cover: Foto ©berggeist007 / pixelio.de

More available books at **www.hansebooks.com**

THE INFLUENCE OF DRESS

IN PRODUCING

THE PHYSICAL DECADENCE OF AMERICAN WOMEN.

J. H. KELLOGG, M. D.,

Battle Creek.

["Annual Address upon Obstetrics and Gynecology," delivered before the Michigan State Medical Society at the Annual Meeting held at Saginaw, June 11 and 12, 1891. Illustrated by a stereopticon.]

As my subject suggests, I am to undertake to show that certain features of the mode of dress common among civilized American women have been, and are, a prominent factor in producing a widespread and marked physical deterioration among the women of this country. Possibly the question may be asked whether such a deterioration exists. It is not probable, however, that it will be worth while to spend any considerable time in attempting to demonstrate the proposition that American women are degenerating physically, before an audience made up chiefly of medical men and women; for has there been a medical convention dinner within the last quarter of a century at which there was not heard the familiar toast, " Woman — God's best gift to man, and the chief support of the doctors"?

A few months ago, I addressed an audience of six or seven hundred young women at an educational center in a neighboring State, upon the subject of physical culture. As my audience seemed to be an amiable one, I ventured to ask a few questions, and among other inquiries, asked how many women present (all of whom had reached adult age) believed themselves to be physically superior to their mothers. A bare half dozen raised their hands, and two or

[3]

three of them timidly looked about, apparently to see if any one present was prepared to contest their claim.

One of the most convincing evidences of the physical failure of American women is to be found in the fact developed by the last census of the United States, that there has been, in the last ten years, an enormous falling off in the birth-rate, as the result of which several million babies are lacking. A lowered birth-rate is a much more serious matter than an increased death-rate, although the immediate result as regards the population might be the same. An increased death-rate may mean nothing more than a temporary increase in the activity of one or more of the causes of disease and death, while a lowered birth-rate means a radical and constitutional fault of some sort, threatening the very existence of the race. Any one who has had an opportunity to become acquainted with the physical condition of the average young woman of the present generation, will be easily convinced that the next census will show a still greater falling off in the birth-rate than the last. A corset-choked woman knows very well that she is quite unfit, physically, for the rearing of children ; and besides the physical unfitness, she finds herself so lacking in fortitude, and so oppressed with nerves and neuralgias and an abnormal susceptibility to pain, that she very naturally shrinks from the physical ordeal, as well as the mental and moral responsibility, which motherhood involves.

Another most significant fact, for which mothers must be held largely responsible, is the enormous business carried on at the present time in the manufacture and sale of infant foods. According to a paper read by Dr. Hoffman, before the American Association for the Advancement of Science, at its last meeting, there is consumed in the United States every year, not less than eight or ten million dollars' worth of infant foods. That these foods are rarely, if ever, perfect substitutes for the child's natural aliment, is well known. What has created such an enormous demand for these substitutes ? Certainly it is not the unnatural increase of the

number of infants which has exhausted the natural food supply ; for I have already mentioned that there has been, in
the last ten years, a falling off in the birth-rate amounting
to several millions.

These evidences point with tremendous emphasis to the
fact of the decline of stamina in American women. A host
of other facts confirming and supporting those given, might
be brought forward ; but I will not thus unnecessarily con-
sume your time, since the proposition is not likely to be
disputed by any intelligent physician who has had wide
opportunities for observation.

But I must not devote more of the half-hour allotted me,
to introductory remarks. 'Fully realizing that I am likely to
incur the displeasure of some of my fair auditors before I
have done with my subject, I may as well declare myself at
once as prepared to defend the proposition that the average
civilized American woman is deformed. This very uncom-
plimentary proposition doubtless impresses my hearers as
somewhat startling. Nevertheless, I believe the evidence
which I shall present will convince the majority of you that,
however repulsive and distressing the fact may be, it is
true.

A penchant for modifying the natural form of the body so
as to produce deformity in some part, seems to prevail quite
extensively in the human race, although it must be admitted
that in many savage, and some civilized tribes, this strange
propensity takes a less dangerous direction than among the
civilized races. The Indian woman of Alaska ornaments her
upper lip with a pin stuck through it. Among the women
of some other savage tribes, fashion demands that a fish-
bone or a piece of wood be inserted in the under lip in a
similar fashion, by means of which the flesh is dragged
down, and a strange deformity produced. The civilized
woman finds the lobe of her ear a more convenient place
from which to hang her jewelry, and so she bores a hole
through this part of her body, and inserts a wire weighted
with a stone, and thus emulates the example of her savage

sisters. There are mothers roaming in the forest, shoeless, hatless, and without other garments than a bark apron and the picturesque designs of the tattooer's pencil, whose solicitude for their children leads them to compress their heads into cones, or to shape them to a fascinating flatness by the steady pressure of a board against the infant skull. Other mothers, less barbarous, but none the less anxious for the welfare of their little ones, squeeze the feet of their daughters into shapeless masses of bone and gristle, in the firm belief that no young lady can make an eligible bride if her foot exceeds in measure the conventional three inches. Still other mothers, more civilized, and none the less fondly thoughtful of their daughters' interests, base their expectations of a successful career for them as much upon the meager dimensions of their waists as upon the comeliness of their countenances or the brilliancy of their accomplishments.

Some years ago, while engaged in some anthropometric studies among Chinese women and the women of the primitive Indian tribes of Arizona and New Mexico, I was forcibly struck, with the marked difference in physical proportion between the savage and the civilized woman. I have made personally, and secured through others, a large number of measurements, which place upon a mathematical basis certain points of difference that are exceedingly pronounced, particularly the larger waist of the savage or semi-civilized woman when compared with the highly civilized woman. I have since extended my studies of the subject to the peasant women of various nationalities, particularly French, German, and Italian women, and a single race of East Indian women. Early in the course of my studies, the thought occurred to me that there might be a positive and constant relation between the external configuration of the body and the mal-position of various internal organs. I accordingly devised a simple apparatus for the purpose of making outline traces of the figure at any desired angle. With this instrument, I have made a large number of tracings (several hundred in all), and have made a careful study of the position of the abdominal and pelvic viscera in each case.

The following is a tabulated statement of some facts which I have collected, and which bear especially upon the matter of waist proportion : —

	Average height.	Average waist.	Average percentage of waist to height.
American women	61.64 in.	24.44 in.	39.6
Telugu women of India	60.49 in.	24.65 in.	40.6
English women (brickmakers who wear heavy skirts)	60.04 in.	25.00 in.	41.3
French women	61.06 in.	28.00 in.	45.4
Chinese women	57.85 in.	26.27 in.	45.4
Yuma women	66.56 in.	36.84 in.	55.2
Civilized men — American	67.96 in.	29.46 in.	43.3
Mrs. Langtry	67.00 in.	26.00 in.	38.8
Venus de Milo	.....	.....	47.6

	Height.	Waist.	Percentage of waist to height.
Average of 43 women, from 18 to 25 years old	60.7 in.	27.1 in.	44.64
Average of 25 women, from 18 to 30 years old wearing corsets or tight bands	62.5 "	23.3 "	37.3
Average of the same 25 women a few months after reforming their mode of dress	62.5 "	27.15 "	43.4
Average of 10 girls, from 9 to 12 years old	....	23.5 "	....
Average of 2,000 men, from 18 to 27 years, measured by Dr. Seaver, of Yale	68.6 "	29.3 "	42.7

A few remarks upon the above figures will render them more significant. Of the 100 American women whose average proportions are given in the table, the majority were upwards of 30 years of age.

Dr. M. Anna Wood, of Wellesley College, has measured 1,100 women between the ages of 19 and 21 years. Her measurements make the height of the average American woman to be 63 inches, waist 24.6 inches ; percentage of waist to height 39.

The Telugu women of India, as I am informed by Miss Cummings, who kindly made a large number of measurements for me, sustain the skirt, which forms almost their only clothing, by means of a cord tied around the waist and drawn as tightly as possible. This is doubtless the reason for the small waists of these women as compared with those of the women of other savage or semi-civilized tribes.

English working women doubtless often do themselves great harm by wearing many heavy skirts attached to waistbands. I once found a young English woman engaged in the very laborious occupation of making brick, kneading the clay with her fist as a baker kneads dough, and beating it into the moulds with her fist, who was at the same time carrying upon her waist the weight of six heavy quilted skirts, with no other support than bands. The average waist measure of a dozen English women brick-makers was 25 inches, and the proportion to height 43.7 per cent.

The German peasant woman, unless she has the misfortune to live sufficiently near some large city to be somewhat influenced by the example of her fashionable sisters, discards waistbands altogether, and wears her garments suspended from the shoulders by means of a waist, which gives her a more vigorous figure than the English peasant woman.

French women are in the last-named respect also more fortunate than their English sisters, by reason of which they enjoy the advantage of a waist proportion of 45.4 per cent of the height.

Chinese women, of whom I have made a large number of measurements, and received much more data through the kindness of Miss Culbertson, of the Home for Chinese Women, San Francisco, and also from a lady medical missionary in China, although considerably below the average height of American women, have two inches greater waist circumference, which is doubtless attributable to the fact that their mode of dress is such as to allow the most perfect freedom of movement and room for development at the middle portion of the trunk.

But the primitive Yuma Indian women of Arizona and New Mexico excel all others whose waist measure I have taken, the average waist proportion being 55.2 of the height.

The famous English beauty, Mrs. Langtry, has recently had published a detailed account of her physical proportions, by which it appears that her height is 67 inches, and her waist measure only 26 inches. Mrs. Langtry takes evident pride in the fact that many of her measurements correspond very closely with those of some ancient Grecian models, but she omitted to call attention to the fact that her waist measure is only 38.8 per cent of her height, while that of the queen of all the ancient statues of women which have been discovered, the famous Venus de Milo, is 47.6 per cent of the height. Mrs. Langtry's waist measure, to be in the same proportion as that of the Greek beauty, should be 32 inches. I have taken the pains to make measurements of a considerable number of male statues, the work of eminent ancient artists preserved in various European galleries, and find the average proportion of the waist to height of seven famous models to be 46.4, or a little less than that of the Venus de Milo.

I have recently made measurements of 43 working women between the ages of 18 and 25 years. These young women were all wearing loose garments, having been induced to do so by a representation of the evils resulting from waist constriction. Some had but recently adopted a healthful style of clothing, while others had enjoyed the advantage of ample waist room for several months or years. In a few instances, corsets and tight waistbands had never been worn. I found the average waist measure of 43 young women, who were selected only with reference to age, to be 27.15 inches, or 44.64 per cent, of the height, nearly 3 inches in excess of the average feminine American waist. The waist of a young woman with this proportion, and of the same height as Mrs. Langtry, would measure 30 inches instead of 26.

Comparative measures made in the cases of 25 of these young women showed that before the adoption of loose garments their average waist measure was 23.3 inches. Since that time there had been an increase in waist proportion to such an extent that the average waist measure at the time the measurements were taken was 27.15 inches. The proportion of waist to height in these 25 young women had increased by the change of dress from 37.3 per cent to 43.4 per cent, and the waist measure had gained 3.85 inches, or 6.16 per cent.

I recently secured the measurements of 10 girls between the ages of 9 and 12 years, and found the average waist measure to be 23.5 inches.

From these facts is it not evident that the small waist of the civilized American woman is a deformity? Can any one assign a physiological reason why the civilized woman should have a smaller waist than the savage woman, or why Mrs. Langtry's waist measure should be 26 inches instead of 32? Certainly no other reason can be given for the abnormal waist of the civilized woman than the fact that this portion of the body has been subjected to abnormal pressure in such a manner as to prevent natural development and to compel the acquirement of a deformity.

If, in answer to the question why the civilized woman of to-day has a smaller waist than the beautiful women of ancient Greece, whose figures furnished models for the sculptors whose masterpieces modern artists have sought in vain to equal, it is said that the change observable is a product of evolution, or a result of civilization, may we not pertinently inquire why a similar change is not to be found in the modern man?

Two other pertinent questions may be asked in this relation :—

1. Why does the civilized woman require a smaller waist than the civilized man? Certainly no physiological reason can be given, and well-known anatomical facts suggest that if there is any natural difference in proportion, woman requires a larger waist than man. She has a larger liver in proportion to her size and weight than man, and the exigencies of motherhood require provision for an increase in waist capacity to which man is not subject. It is interesting to note, also, in this connection, that the waist proportion of the Venus de Milo,

who may be considered as the typical woman of the ancient Greeks, is 47.7 per cent, while that of the average Grecian man, already shown, is 46.4.

We can draw but one conclusion from these considerations, namely, that the small waists of the women of modern times are an abnormality. My tables also show the average modern feminine waist to be nearly two per cent larger in proportion to the height than the modern male waist, when it is allowed a chance for natural development.

2. A second question to which I invite attention is, Why does the waist of the civilized woman cease to grow at the age of 10 or 12 years, while the rest of the body continues to develop? Lungs, liver, stomach, spleen, bowels, pancreas—all the organs which occupy the region of the waist line, continue to grow, but the waist of the civilized woman absolutely refuses to increase in size, notwithstanding the developing force beneath it, after the age of 12 or 14 years. I find the average waist measure of girls from 9 to 12 years of age to be 23.5. I have in some instances found the waist measure in girls of 12 to be 26 inches. The rational answer to this question is the fact that about this age the constricting influence of tight bands, corset waists, or corsets begins. The fashionable dressmaker insists that the young lady's figure must be "*formed*," and so as she develops, she grows into a mould like a cucumber in a bottle. And thus it happens that we find the civilized woman with a waist disproportionately small, as we find, among the aristocratic class of Chinese women, dwarfed and misshapen feet. The small-footed woman of China, in consequence of her deformity, is compelled to hobble about in a most ungraceful fashion, requiring usually one or more persons to sustain her in keeping her balance. She cannot run, skip, or dance as can her large-footed sisters. She is willing, however, to endure the inconveniences of being a cripple and the loss of the use of her feet and legs rather than forego the pleasure of being in fashion. If the sacrifices which the civilized woman makes to fashion were no greater, there would be comparatively small ground for complaint, but the

constant girding of the waist results in mischiefs of vastly greater magnitude than those which the Chinese woman inflicts upon herself.

As the flat-headed woman watches with interest and growing pride the progressive depression of her infant's skull, while from day to day she binds more tightly upon it the flattened disc of wood ; and as the Chinese woman glories in the shriveled and misshaped stump of what was once her child's foot, as a developing mark of aristocratic gentility, in like manner does the civilized mother pride herself on the smallness and roundness of her daughter's corset-deformed waist, disregarding alike the suggestions of art, the warnings of science, and the admonitions which nature gives in the discomfort and distress occasioned by the effort to secure a change in the natural contour of the human form which is more monstrous in its violation of the laws of beauty, more widely at variance with the dictates of reason, and more disastrous in its consequences to bodily health and vigor, than any similar barbarity practiced upon themselves or their children by the members of any savage or semi-savage tribe.   How such a disfigurement of the physique could ever have come to be considered desirable or beautiful, is a problem hard to solve, since it involves not only an enormous loss of strength and vigor, but a violation of all the relevant precepts and principles of art which have been handed down to us by the great masters, as well as rules of hygiene in which all medical men of every age agree.

I may ask further, How does it happen that the waist of the average girl of 9 or 12 years measures 23.5 inches, while the waist of the young woman of from 18 to 30 years who has worn corsets or tight bands for a number of years, is only 23.3 inches ?   Why should the waist *decrease* in size with age while every other bodily dimension *increases* ?

Still another question of interest arises from the fact to which almost every woman can testify, that the waist of the average woman accustomed to constriction from clothing, increases in measure whenever it has an opportunity for development, as when the common mode of dress is exchanged for a more healthful one, or ordinary clothing laid aside for a few weeks, as during confinement to bed from illness. Probably few women will question the fact that the waist is made smaller by constriction of the corset and tight bands. A lady said to one of my nurses, when she learned of her healthful mode of dress: " But how do you manage to keep your stomach down?" The corset is worn with a deliberate purpose of modifying the form of the waist, which it does to the great damage of health and vigor. I have shown by careful measurements in some hundreds of cases, that the waist of an adult woman increases within a few months, under the influence of proper clothing and proper exercise, from one or two to six or seven inches.

Let me call your attention more directly to some of the important particulars in which the ordinary mode of dress among civilized women, especially constriction of the waist, results in physical injury. The chief of these are : —

1. Downward displacement of all the abdominal and pelvic organs, and numerous functional and organic diseases growing out of this disturbance of the static relation of these organs.

2. Lack of development of the muscles of the trunk, which by long compression and disuse, to a very large degree lose their functional activity, resulting in relaxation of the abdominal walls, weakness of the muscles of the back, general physical feebleness, and destruction of the natural curves of the body, which are not only necessary for health, but also essential to physical grace and beauty, and the development of many bodily deformities, such as drooping shoulders, flat · or hollow chest, sunken epigastrium, straight spine.

3. An ungraceful and unnatural carriage of the body, in sitting, standing, and walking.

4. An abnormal mode of respiration.

The idea that a displaced stomach can be a possible cause of disease or inconvenience may be new to some. Nevertheless, the researches of Glenard, Bouchard, Dujardin-Beaumetz, and other eminent French physicians, have shown beyond room for doubt that displacement of the stomach, bowels, kidneys, liver, and other abdominal viscera, may be productive of the most pronounced disturbance of health and a source of great inconvenience. Indeed, from my own studies on this subject I have become convinced that a displaced and dilated stomach is more likely to be productive of immediate and harmful consequences of a grave character, than displacement of the pelvic viscera. But before one can fully understand the relation of waist constriction to displacement of the abdominal viscera, it will be necessary to call to mind a few important anatomical facts.

The trunk is practically divided into two cavities. The division of the lower cavity into pelvis and abdomen is an artificial and not an anatomical subdivision, useful for the purpose of description, but misleading and confusing, unless ignored in studies concerning causation and pathological relations. Anatomically, the trunk is divided by the diaphragm into two cavities only, the upper containing the chief organs of respiration and circulation, and the lower containing the principal organs of digestion and the genito-urinary apparatus. The chief anatomical facts which I desire to be kept in mind are, the normal position of each of the viscera which occupy the lower cavity of the trunk, and the mode in which these various organs are held in place. It will be remembered that the liver, spleen, pancreas, and stomach are all located above or at the waist, as shown in the accompanying diagram after Ziemssen. Plate II. The transverse colon lies at the waist line, the point of junction of the ascending and transverse colon on the right side dropping a little below the line, while the point of conjunction of the transverse with the ascending colon at the left side rises considerably above the waist line, being held in place by the pleuro-colic fold of the meso-colon. The kidneys lie just at the waist. The greater portion of the space below the waist is occupied by the small intestines, the bladder, and the rectum, with the uterus and its appendages in the female, and the prostate gland and other special structures in the male. It is notice-able that the organs of the greatest weight and functional importance are located at or above the waist.

How are all these important organs held in position? Although fitted together with the nicety of an articulation, the viscera are certainly not held together by anything cor-responding to the firm ligamentous bands which unite the osseous elements of a joint. Every abdominal surgeon will testify to the extreme propensity for escaping from the ab-dominal cavity when the slightest opportunity offers, mani-fested by some of the viscera. The so-called ligaments which hold in place the liver, stomach, spleen, and bowels, cannot properly be called ligaments, as very little ligament-ous structure enters into their composition. The same must be said of the ligaments which are supposed to support in place the uterus and ovaries, although it must be added that some of the uterine ligaments contain muscular tissues which play a very important part in maintaining the uterus in its proper relation to the trunk and the contiguous organs. I think the idea is gaining ground among those who have made a special study of this subject, that the chief factors in the support of the pelvic viscera, as well as other of the organs of the lower trunk cavity, are the tone of the muscular walls of the abdomen and the juxtaposition of the organs themselves.

Compression of the waist necessarily involves displace-ment of the organs occupying this portion of the trunk. The unyielding character of the chest walls, and the resist-ance of the diaphragm prevent any considerable displace-ment in an upward direction. Consequently, the necessary result of waist-compression, either by the corset or by tight bands, is, that the liver, stomach, bowels, and other organs occupying this zone of the body, are carried downward. The same compressing force which diminishes the circum-ference of the body at the waist, interferes with the normal activity and development of the muscles which form the anterior wall of the lower trunk, so that they offer little resistance to the displacing force applied at the waist.

In nearly twenty years of medical practice, I have had to deal almost exclusively with chronic disorders of various sorts, and especially with two classes of chronic disease,— digestive disorders, and maladies peculiar to women. Having under observation from 1,000 to 1,500 cases annually, under conditions favorable for careful study and comparison, I long ago noticed the remarkable frequency of the association of certain forms of pelvic disorder (especially in women with a narrow waist and a protruding abdomen). I did not, however, attach so great importance to the matter as I should have done, I frankly confess, had I not had an erroneous notion respecting the normal contour of the female figure. It was only after careful study of this matter among savage women, and women whose figure had never been modified by the deforming influence of the ordinary civilized dress, that I acquired a basis from which to view this subject in a rational way. I then began careful inquiry into the matter, and for several years back have made, in all cases of pelvic diseases of women coming under my observation, a careful study of the condition and relative position of the various abdominal viscera, as well as of the pelvic organs.

In 250 cases of women suffering from pelvic diseases, taken consecutively and without selection, in each of which a careful examination was made with reference to the condition and position of each of the abdominal viscera as well as of the pelvic organs, I observed the following disturbances of the static relations of the viscera :—

In 232 cases, downward displacement of stomach and bowels.

In 71 cases, right kidney distinctly movable and sensitive to pressure.

In 6 cases, both kidneys freely movable.

In 9 cases, downward displacement of the spleen.

In one of these cases, the spleen lay at the bottom of the abdominal cavity. I have made a large number of outline tracings in cases of women suffering from pelvic diseases, and supplemented these by careful examination of the position and conditions of the abdominal and pelvic viscera, with the following results, as regards the relation of the static changes in the abdominal organs, to similar changes in the organs of the pelvis.

In 150 cases of pelvic disease, the stomach and bowels were displaced in 138 cases.

In 66 cases, the stomach and bowels were displaced without displacement of the uterus. In 26 of these cases, there was also a displacement of one kidney, and in five, a displacement of the liver.

In only seven cases was there displacement of the uterus without displacement of the abdominal viscera, and three of these were cases of large uterine fibroids in which the visceral displacement was probably present, but masked by the morbid growth.

I shall have thrown upon the screen, presently, outline tracings of the figures of some of these cases, which will show very clearly the amount of visceral displacement occasioned by an improper dress. My statistics seem to show very clearly that visceral displacement is not a disease which is especially confined to the pelvis. Indeed, a careful study of the means by which the pelvic organs are held in place, suggests that they are better provided for in this respect than any other of the viscera below the diaphragm. The data which I have collected respecting the relative frequency in the displacement of the pelvic organs, and other organs of the abdominal cavity, clearly support this idea. In 150 cases of pelvic diseases, there were only four cases in which displacement of the pelvic organs was present without displacement of one or more of the abdominal viscera, while there were 66 cases in which the stomach and bowels were displaced without any displacement of the pelvic organs. In 26 of these cases there was also a displacement of the kidney, and in five a displacement of the liver. It is evident, then, that visceral displacement of the organs of the lower trunk must be regarded (of course

2

leaving room for exceptions) as a general disorder, affecting more or less the entire contents of the abdomen and pelvis, rather than as a disease confined to one or two of the organs in which the subjective symptoms happen to be most prominently manifested.

How a displacement of the stomach, a kidney, the bowels, the uterus, or an ovary, may occasion disease, is a pathological question which it is not necessary to spend time in discussing, since the disturbance in blood-circulation, and hence in nutritive changes (possibly, also, in the supply of nervous energy), and the development of abnormal and pernicious nerve-reflexes, are etiological factors, the influence of which is too well known and understood to be disputed, and which are likely to come into active operation under the morbid conditions established in an organ crowded by abnormal pressure out of its proper place. Nature has placed each internal organ in the position in which it can do its work most easily and efficiently; and the studies of the results of visceral displacement which have been made by eminent scientific physicians, have shown that to morbid conditions of this sort may be fairly attributable the most serious, and not infrequently the most obstinate, disturbances of some of the most important vital functions, and through them, of all the other functions of the body.

The question may arise, whether we are treating the subject fairly, in charging upon errors in dress, so great and so serious modifications of the human form as we have pointed out, and whether it is not possible that visceral displacements in some of those cases to which I have called attention are to be found in men as well as in women. In order to place this subject upon a rational basis, I recently made a careful examination respecting the position of the stomach, liver, and bowels in 50 working men and 71 working women, all of whom were in ordinary health.

In the 71 women examined, prolapsus of the stomach and bowels was found in 56 cases. In 19 of these cases, the right kidney was found prolapsed, and in one case, both kidneys. The 15 cases in which the stomach and bowels were not prolapsed were all persons under 24 years of age. None of these had ever laced tightly, and four had never worn corsets or tight waistbands, having always worn clothing suspended from the shoulders. It is noticeable that in a number of cases in which corsets had never been worn, tight waistbands had produced very extensive displacement of the stomach, bowels, and kidney. In one of these the liver was displaced downward.

In the 50 men, I found only six in whom the stomach and bowels could be said to be prolapsed. In one the right kidney was prolapsed. In only three was the degree of prolapse anything at all comparable with that observed in the women, and in these three (and in one other of these six cases, making four in all) it was found on inquiry that a belt or something equivalent had been worn in three cases, as a means of sustaining the pantaloons. In one case the patient attributed his condition to the wearing of a truss furnished with a belt drawn tightly about the waist. This belt had been worn a sufficiently long time to be an ample cause for the visceral displacement observed. In the two cases of slight visceral prolapse in which belts had been worn, there was considerable deformity of the figure due to general weakness, and a habitual standing with the weight upon one foot. By comparison, we see the relative frequency of visceral prolapse in the men and women examined, was 12 per cent of the men and 80 per cent of the women. In other words, visceral prolapse was found to be 6⅔ times as frequent in women as in men. It is also noticeable that, with the exception of two cases of visceral prolapse in the men, the visceral prolapse in the men was due to the same cause which caused visceral prolapse in women ; viz., constriction of the waist. It makes no difference, of course, whether the constriction is applied by means of a corset or a waistband or a belt.

I have met a number of cases of visceral prolapse in men in which the disease was directly traceable to the wearing of

a belt. One case was that of a military officer, who wore a
tight sword belt, in which he carried almost constantly a
heavy sword. I have also made some observations of the
same character among blacksmiths, who have a habit of sustaining their pantaloons by means of the strings of their
leather aprons tied tightly about the waist, the suspenders
being loosened so as to give greater freedom to the movements of the arms. Farmers, also, sometimes seek to liber-
ate their shoulders by wearing the suspenders tied about the
waist. Leaving out of consideration the four cases of men
in whom the visceral displacement was due to the same
causes which produce this morbid condition in women, we
find but two cases in which the viscera had become displaced from other causes, or one in twenty-five,— a frequency
just one twentieth of that in which this diseased condition is
found in women who consider themselves enjoying ordinary
health.

These facts; it seems to me, are amply sufficient to estab-
lish my proposition,—that constriction of the waist is the
cause of downward displacement of the pelvic viscera, and
of the diseases which naturally grow out of such disturbances
of the static relations of the organs occupying this portion of
the trunk.

The injury inflicted upon the body at its central portion
by constriction of the waist, attacks the very citadel of its
strength and vigor, the stomach and its associate organs con-
stituting the headquarters for the supply of force and energy
for the whole system. It is doubtless for this reason that the
great abdominal brain, the largest collection of nerve matter
in the sympathetic system, is found in such close relation to
the stomach. Lying, as it does, exactly in the plane of the
waist, any abnormal pressure at this point must act directly
upon this great center of reflex nervous activity.

By the inactivity of the muscles of the trunk, and the failure
of development due to continued pressure, the muscles of the
central and anterior portions of the trunk become abnormally
weak, so that their natural tone is insufficient to support the

abdominal contents in their normal position. As I have already shown, an additional injury results from the failure of these weakened muscles to perform their duty as guys, which balance the upper half of the pelvis upon the trunk, and by their efficient action in health, maintain a graceful and healthful poise of the body.

The strong and beautiful curves which are observed in a spirited horse are not only attractive from an æsthetic point of view, but are also of the highest significance from a physiological standpoint. In the healthy, vigorous animal one observes that the head is held high, the neck and back strongly curved, the limbs firmly set, and the whole expression indicates vigor and strength. The same is equally true of the human body. An erect head, well curved back, prominent chest, retracted abdomen, and firmly set limbs, are indicative of an energized carriage of the body which is characteristic of health. The flat chest, posterior dorsal curve, projecting chin, protruding abdomen, are equally indicative of a relaxed and weak carriage of the body, characteristic of feebleness and disease. The spiritless and tired horse does not *hold* his head down ; he lacks the vigor and disposition *to hold it up*. So the woman who has been accustomed to the support of stays of steel or bone, finds herself, when without these means of support, feeling, as she says, " as though she would fall to pieces." The muscles of the waist lack the ability to balance the chest and shoulders upon the hips.

As I shall show you presently in the outlines which will be thrown upon the screen, the direct effect of the corset, and of any constriction of the waist, is to break down the natural curves of the back, straightening the spine, thus depressing the chest, and causing the shoulders to fall forward, and producing general collapse of the front wall of the trunk.

In consequence of the weakening of the muscles which support the trunk, and especially weakness of the waist muscles, an ungraceful and unnatural carriage of the body appears, not only in walking and standing, but in sitting. The weak-waisted woman is comfortable only when sitting in a rocking or easy chair. She cannot be comfortable unless the back is supported ; consequently, in sitting the muscles of the trunk are completely relaxed, thus causing collapse of the waist and protrusion of the lower abdomen by the depression at the waist occasioned by the depression of the ribs.

Such persons, in standing, assume a great variety of awkward and unhealthful positions, some of the most common of which will be shown presently upon the screen. The most common faults are dropping the shoulders, projecting the chin, hips too far forward, weight resting upon the heels or upon one foot, and a general lack of even and graceful balance of the body. In walking, the forward position of the hips makes it impossible to plant the whole sole of the foot down at once and firmly, so the weight is thrust continually upon the heels. This difficulty is increased by wearing high-heeled shoes. A swinging, swaying, wriggling, or otherwise awkward gait, is the most common mode of walking one sees in women, very few of whom are good walkers, in consequence of the inability to balance the body, through the weakness of the muscles of the waist.

The fourth charge which I have made against the common mode of dress, in which the waist is constricted, is that it induces and necessitates an abnormal mode of respiration.

In normal breathing, the shape of the chest-cavity is changed in the act of inspiration in such a manner that its diameter is increased in all directions. The greatest increase, however, is in its longitudinal diameter, due to flattening of the diaphragm ; and in the lateral transverse diameter of the lower part of the chest, due to the action of the inspiratory muscles, and, according to Brüger, also in part due to the depression of the abdominal viscera by the contracting diaphragm. In normal respiration in children of both sexes, and in both men and women of savage tribes, in whom the dress of the two sexes is practically alike, the chief movements noticeable to the eye in inspiration are widening of the chest at its lower part and bulging of the abdominal wall. There is at the same time a rhythmical action of the muscles of the pelvic floor, induced by the increase of abdominal pressure resulting from the flattening of the diaphragm, acting against the resistance of the tense abdominal muscles.

That the respiratory movements are practically alike in adult persons of the two sexes, I think has been fully established by the observations of Mays, Dickinson, and others, as well as by my own studies upon Indian women of various tribes, Chinese women, Italian peasant women, and American women whose breathing has never been interfered with by tight-fitting clothing.

The relation of corsets and tight bands to respiration has usually been studied with reference to their influence upon the lungs or the respiratory process. The important relation of the respiratory process to the abdominal and pelvic viscera has too often been overlooked, although the disturbance of the normal relation existing between respiration, and the circulation of the blood in the abdominal and pelvic viscera, is undoubtedly a matter of far greater importance than any interference with the respiratory process occasioned by constriction of the waist. The effect of inspiration is to increase abdominal tension. This is accomplished by the flattening of the diaphragm, which is facilitated by the increase in the lateral transverse diameter of the lower part of the chest, induced by contraction of the serratus and other inspiratory muscles. The effect of abdominal tension is to facilitate the emptying of the veins of the portal circulation, in which there is a natural tendency to congestion, as the result of the resistance of the hepatic capillary system, which intervenes between them and the general venous system. In normal respiration, in which the intra-thoracic pressure is diminished by proper expansion of the chest cavity, this emptying of the portal circulation is also facilitated by a sort of suction action, which draws the blood from the abdominal viscera into the thoracic cavity. Thus, in normal respiration there is a double action, the tendency of which is to accelerate the circulation in the abdominal and pelvic organs ; and it is reasonable to suppose that the health of these organs must largely depend upon a continuous and efficient action of this pumping process, which is so essential a feature in the maintenance of the blood current in this region of the body.

When the waist is constricted, both elements of the respiratory process through which the abdominal pelvic circulation is assisted, are seriously weakened. The increase of tne abdominal tension, resulting from the pressure of the diaphragm, is prevented by the fact that the transverse diameter of the lower portion of the chest is not only diminished, but fixed. The lateral attachments of the diaphragm are thus approached in such a manner that this muscle is rendered incapable of efficient contraction. At the same time, the intra-thoracic negative pressure is diminished through the crippling of the inspiratory act. The lower portion of the chest being held firmly, any increase in the transverse diameter of this part is impossible. The normal descent of the diaphragm being prevented, the longitudinal diameter of the chest cannot be increased to the proper extent. The chest is left free to act only in its upper part, the elasticity of which is much less than that of the lower portion, in consequence of the rigid character of the ribs, and the shortness of the cartilages which connect the ribs to the sternum, as well as the comparative weakness of the muscles which act upon this portion of the chest.

The crippled condition of respiration in a woman whose waist is constricted by a corset or tight bands, is clearly shown by the readiness with which such a woman gets out of breath when called upon to make unusual exertion, or when there is a special demand for lung activity from any other cause. The first thing done for a fainting woman is to cut her waistbands and corset strings ; but no one would ever think of tearing open a man's vest or slitting up his shirt front under the same circumstances.

The proper action of the chest may be aptly compared to that of a pair of bellows. The lower ribs, to which the strong breathing muscles are attached, serve as the handles.

The breathing apparatus of a woman whose waist is constricted by a corset or tight bands, is nearly as much embarrassed in its action as would be a pair of bellows with the handles tied together. The clavicular respiration, so conspicuous in women who constrict the waist, is not seen among savage women, nor in a woman whose respiratory organs have not been restricted in their action by improper clothing. That this mode of breathing is quite abnormal might be fairly inferred from the structure of the upper part of the chest, which is certainly not such as to suggest any considerable degree of mobility. But this mode of breathing is not only abnormal, but, as I think I have already shown, it may be productive of disease. This is true of ordinary respiration, but it is most emphatically true of forced respiration, such as is induced by singing or active muscular exercise. Under the imperative demand for an increased supply of air, the respiratory muscles are made to act with undue violence. In consequence of the constriction and the compression of the abdominal walls by the corset, this abnormal force is largely expended upon the organs of the pelvis, which are forced down out of position. The pelvic floor is more yielding than the rigid walls of the upper chest, and is depressed, thus laying the foundation for chronic displacement. A civilized woman, wearing the common dress, cannot expand her waist more than one fourth of an inch when taking a deep inspiration. Expansion must occur somewhere, and the abnormal mode of dress necessitates that it shall be at the upper and lower extremities of the trunk. The greater resistance of the upper ribs, and the yielding character of the structures which form the pelvic floor, lead to a lowering of all the organs which are dependent upon the latter for support.

The tracings which I shall present also show an important fact as to the influence of constriction of the waist upon breathing. These tracings were made with a pneumograph, the tracings obtained by which represent the whole of the respiratory movement. Fig. 1, Plate I., represents normal respiration. Noting the time relation between inspiration and expiration, it will be observed that expiration is perceptibly longer than the movement of inspiration. I find this relation to be, on the average, about five for inspiration and seven for expiration. Fig. 2 is a fac-simile of the tracings produced by the same person while wearing a corset, who without a corset produced the tracings of Fig. 1. It will be seen that there is an increase in the time of inspiration as compared with expiration, which one would naturally expect from the resistance offered by the corset. It will also be noticed that a marked change in the form of the tracings is produced by the constriction of the waist. The expiratory portion of the tracing, which appears above the horizontal line, drops suddenly, instead of making a gradual decline, as in normal respiration. The tracings obtained from the woman in a corset, show most clearly an abnormal resistance to inspiratory action.

In natural breathing, the action is chiefly at the waist, although the entire trunk wall and every organ within the trunk participates in the movement. The action begins with expansion, first at the sides, and then in front, then a slight elevation of the upper chest, and, in forced respiration, a slight drawing in of the lower abdomen at the same time. In ordinary respiration, there is simply a lifting forward of the whole front wall of the chest and abdomen, the movement extending all along the line from the upper end of the breast bone to the pubis.

The so-called abdominal respiration is unnatural and unhealthful; indeed, it has been in many cases productive of serious injury. Teachers of elocution and vocal music often instruct their pupils to breathe abdominally; that is, to give prominence to the movements of the lower abdomen in breathing. When the waist is constricted, the inability of the chest to expand at the sides, compels an exaggerated movement downward, so that the viscera are forced down into the abdomen to an unusual extent. In natural respiration, the expansion of the waist, or increase in circumference of the

trunk at its center, prevents this excessive downward movement. It will be readily seen how by violent efforts to force
the breath downward with the waist confined so as to prevent
proper expansion, the supporting ligaments of the various
viscera might in time be so stretched as to produce a general
sag of the abdominal contents.

Correct breathing is as necessary to the health of the
pelvic and abdominal viscera as to a healthy condition of
the lungs ; for the respiratory act not only pumps air in and
out of the body, but draws blood to the heart, assisting particularly the portal circulation, and thus also aiding in the
absorption of the products of digestion, and so facilitating
the digestive process. It is quite possible also, that the
rhythmical movements imparted to all the viscera of the trunk
by normal respiration, are a sort of vital gymnastics, essential to the health of each organ.

It is thus evident that, in its interference with the proper
respiration, as well as from the mechanical injuries which it
inflicts, the common mode of dress, which involves constriction of the waist, is the most potent means of impairing the
health and vigor of the whole body, and may justly be reckoned as perhaps the greatest of all factors in the general
decadence in physical vigor so apparent among women of
the present and rising generation.

That there has not been a general rebellion against this
unnatural and mischief-making mode of dress on the part of
the intelligent women of this enlightened age, is probably
due to the popular but fallacious idea which seems to be so
thoroughly fixed in the minds of both men and women, that
woman is "the weaker vessel," and naturally subject to ailments and weaknesses and general physical inefficiency from
which men enjoy immunity. Any one, who has made himself familiar with the activity of the women of savage nations,
or even the women of the peasant classes in civilized countries, must have recognized the fallaciousness of this popular
idea, which had its birth in the age of chivalry, and has come
down to us along with numerous other fancies and supersti-

tions, which have no foundation either in natural experience
or physiological science.

The average civilized woman is certainly very much in-
ferior to the average civilized man in physical vigor. The
constancy of this observation has led both the profession and
the laity to regard women as naturally weaker than men.
But that this is not necessarily so, is shown by the constant
experience and observation of travelers among uncivilized
tribes. Travelers in China are often astonished at the immense loads which Chinese women carry upon their shoulders. Some years ago, I saw a woman in the market-place at
Naples, Italy, carrying off upon her head an immense load of
vegetables, which required two men to lift it into position.
De Saussure relates that when he had finished his observations in the valley of Zermatt, he packed a collection of min-
eralogical specimens in a box, and called for a porter to carry
it out of the valley, as the mountain roads were too steep to
be traveled by four-footed animals of any sort. After a
fruitless search for a man who was able to transport his box
of specimens, he was finally told if he wished a porter to
carry his package he must employ a woman, as no man could
be found who was able to even lift the box. He accordingly
engaged a woman who offered herself for the service, and
stated that she carried the heavy box of minerals over the
steep mountain roads without the slightest injury either to it
or to herself. Stanley reports that the two hundred women
porters whom he employed on one of his expeditions, proved
to be the best porters he ever had in Africa.

When in England, a few years ago, I made an expedition
into the "black country" for the purpose of studying the
women brick-makers and nail-makers of that region. I
found at Lye, some of the finest specimens of well-developed women I ever saw anywhere, women who had spent
all their lives in brick-yards or before the forge, swinging
the blacksmith's hammer and making the anvil ring. These
women never go in out of the rain for fear they will get wet
and take cold, and although working in mud and water a

great share of the time, have no other protection for their feet than shoes, often full of holes and almost without soles, and wholly inadequate to protect the feet from water. They are constantly engaged in lifting heavy weights. One woman I saw tossing and kneading upon a block a mass of clay, which, as I found by actual test, weighed over sixty pounds. She handled it in her hands as though it were only a small mass of dough ; and although thus employed from early morn until late at night, she was in no way disabled by her occupation. A physician of long experience, who practiced in the place, assured me that so far as his practice among women was concerned, it amounted to nothing in a gyneco-logical way, but that his obstetrical practice was very large. Not long ago, a public meeting was held in Birmingham, England, by the nail-makers of that district, for the purpose of protesting against the employment of women in the business of nail-making. The reason given by a prominent member of the association for this objection to the employment of women, was that by this kind of labor a woman became so "unsexed" that she could outwork a man, continuing her labor hours after a man was completely used up.

These facts, and many others which might be cited, show that woman is not necessarily weaker than man. The weakness of woman is not due to natural constitution, but to a vicious mode of dress and neglect of physical exercise, although, possibly, heredity has some influence in the matter at the present day.

The practical bearings of this question are too evident to require more than mention.

1. It is evident that pelvic disease involving the displacement of organs is only a part of a general disorder in which every organ below the diaphragm may be involved, and any system of treatment which addresses itself exclusively to the disorders found present in the pelvis, must be unsuccessful. Here is to be found the secret of the failure of so many methods and systems which have been proposed for the relief or cure of pelvic disease, particularly displacements.

I do not hesitate to make the assertion that any method of treatment, either medical or surgical, which does not address itself to the removal of the causes of the disorder as well as to its effects and amelioration of symptoms, must result in failure. Temporary relief, often apparent cure, may be effected, but sooner or later the patient will find himself in the same wretched condition as before. This explains the almost universal failure of pessaries, local application of electricity, operations upon the perineum, and the various forms of anterior and posterior colporrhaphy, operations for shortening the round ligaments, ventro-fixation of the fundus, and a great variety of other methods and procedures which have been adopted for the relief of the various forms of displacements of the pelvic organs. The pessary sometimes succeeds, provided there is some coincident change in the habits of the patient which increases the strength of the muscles of the waist and abdomen. But, in my estimation, nothing can be more absurd than to thrust a pessary up among a mass of prolapsed abdominal and pelvic viscera, stretched away from their normal moorings, jostling one another about in the abdominal cavity, swaying in whatever direction the body happens to incline. No wonder that such patients often complain that the pessary gives pain. Certainly it is no marvel that ulceration, ovarian irritation and inflammation, and even salpingitis, are not uncommon results.

Thirty years ago, Banning undertook to effect a cure of pelvic disorders by means of braces which supported the trunk in a natural position. The weak point in this system was its inability to give strength to the weakened muscles. An external skeleton consisting of an iron frame-work is no more efficient in developing the muscles of the trunk than one composed of hickory or whalebone stays. The "Neptune's girdle," or "*umschlag*" of the old German water-cures, not infrequently perfected cures by allaying local congestions, irritation of the abdominal sympathetic ganglia, and especially by supporting the relaxed abdominal walls, and holding up in position the prolapsed viscera. Patients are sometimes cured by being sent on long journeys abroad, in which they gain muscular strength and vigor by mountain-climbing, horseback-riding, and the active exercise necessarily involved in sight-seeing.

Cures have been effected by each of these and other haphazard methods of treatment, but they were accidental, and not due to scientific methods, and patients were not infrequently made worse. I have known of cases in which young women were injured for life by being advised by their physicians to exercise in a gymnasium, without the same careful prescription as to the kind and amount of exercise to be taken, as a judicious physician would give respecting the administration of a powerful drug.

2. It must be evident that a large share of the symptoms present in cases belonging to the class which is generally referred to the gynecologist, are really due to disorders of other organs which are involved in the general disturbance, or, as the French call it, *desequilibration*, of the viscera of the lower half of the trunk.

Most women suffering from pelvic diseases complain of pain when on the feet, dragging pain in the bowels and the lower portion of the back, pain at the extreme lower end of the spine, soreness and pain in the region of the navel, a feeling of lack of support in the lower abdomen, a sensation commonly described as goneness at the pit of the stomach, weakness of the lower limbs, pain in the back, crawling, tingling, numbness, stinging, and other sensations in the legs, cold hands and feet, burning of the soles and palms. Sometimes the patient says she is only comfortable, when on her feet, when holding up the bowels with the hands. Such patients tenaciously cling to the corset, because they evidently need some support. These patients also often complain that when they undertake to stand without a corset, there is such a sinking at the stomach that they are compelled to sit down. The evident cause is the dragging of the prolapsed bowels and stomach, occasioned by the relaxation of the abdominal muscles by which the branches of the pneumogastric and sympathetic nerves are put under an unnatural strain. It is evident that in these cases a large part of the symptoms are due, not to the pelvic disorder, but to the general disease of which this is a part.

If we expect to cure a woman who is a chronic sufferer from pelvic disorders, we must treat the patient rather than the malady. This is a principle which applies, in fact, to most chronic disorders ; and a failure to recognize this principle is the rock upon which professional effort often splits. It is as hopeless to undertake to cure such maladies by the usual routine methods, which are addressed to local symptoms and conditions only, as to expect to kill a noxious weed by picking off its flowers or a few of its leaves. The whole disease must be eradicated, root and branch. This can only be accomplished by the removal of all the morbid conditions which are the real causes of the multitudinous symptoms by which the disorder is recognized and for which it is often named. Rational treatment of this class of diseases must, then, include, first of all, the adoption of a proper dress, which will be one in which every muscle of the trunk will have perfect freedom to act. The patient must be instructed to have her dress measure taken with the waist fully expanded, and to allow an inch or two for growth, in the hope that to some degree she may overcome the deformed condition which she has induced by ignorant obedience to fashion, rather than to the laws of physiology and the dictates of common sense.

Health corsets are a device of the devil to keep women in bondage who are seeking for deliverance from the weakness and misery from which a really healthful mode of dress might emancipate her. Shoulder braces and harnesses of every description are, on the whole, a snare and a delusion. The only correct principle is to suspend everything from the shoulders by means of a waist which will equally distribute the weight upon natural bearings, and at the same time give latitude for the greatest freedom of waist movement.

If all women would at once adopt a healthful mode of dress, probably half of our profession would be obliged to seek some other calling. Certainly, at the present time, more than half our business consists in efforts to repair damages which ignorant women have inflicted upon themselves. Neither a proper knowledge of the requirements of the body, nor a just consideration of the principles of beauty, justifies the popular mode of dress. The idea that a small waist or a round waist is beautiful, is a mischievous and dangerous notion, which ought to be eradicated from the public mind. Nature never made a waist round, slight, or tapering, as though it were chiseled out of a block of wood; and why should we allow ourselves to be persuaded by the fashion-mongers that a thing which from an artistic standpoint is truly hideous, is otherwise than monstrous and repulsive ? An artist who would make a nude figure with the waist modeled after a French corset, would not be allowed to exhibit his work in any respectable gallery.

A singular illustration of the inconsistency of human nature is to be found in the fact that the same artist who takes so great care of his "model's" figure that he will not allow her to wear a corset, or subject herself to waist constriction of any sort, never thinks to criticise his wife, who squeezes herself into a French mould of the latest pattern, regardless of the fact that the circumference of her trunk is decreased by several inches at the middle, only at the expense of a commensurate increase below the waist, making an unsightly protuberance of displaced adipose tissue, relaxed abdominal muscles, and a promiscuous assemblage of stomach, bowels, kidneys, spleen, and other things, which have been forcibly ejected from the snug corners in which nature carefully stowed them away, and thrust into an unnatural and unsightly mass below. We see in the enormous busts and bustles which fashion prescribes, an evident attempt to cover up the uncouthness of form which the corset and other fashionable modes of torture have induced by means of these excrescences, and by their aid to approach as far as possible to the ideal figure, which, in its native grace and beauty, requires no such accessories.

EXPLANATION OF PLATES.

PLATE I.

FIGURE 1. Pneographic tracing, showing the respiratory movements of a healthy woman. That portion of the curve above the base line represents *expiration ;* the curve below the line represents *inspiration.* This tracing was obtained by means of a new form of pneumograph, or pneograph, which represents the whole respiratory movement, and which I have elsewhere described.

FIG. 2. Pneographic tracing furnished by a woman wearing a corset. The subject was the same person who furnished the preceding tracing, and the tracing was made with the same instrument adjusted in the same manner. The evident increase in the length of the expiratory movement, or rather decrease in the length of the inspiratory movement, as well as the change in form of the expiratory movement, are strongly suggestive of the interference with respiration occasioned by constriction of the waist.

FIG. 3. Pneumographic tracing of the upper and lower costal movements in breathing, furnished by a healthy man. The pneumograph employed in taking this tracing was a modified form of the instrument designed by Paul Bert. In obtaining the upper costal movement, the instrument was adjusted at the middle of the sternum and the spine opposite. In obtaining the lower costal tracing, the instrument was adjusted to the sides of the chest.

FIG. 4. Pneumographic tracing furnished by a civilized woman wearing a corset. It will be noticed that this tracing is almost exactly the reverse of the preceding.

FIG. 1. Pneographic tracing of a healthy woman.

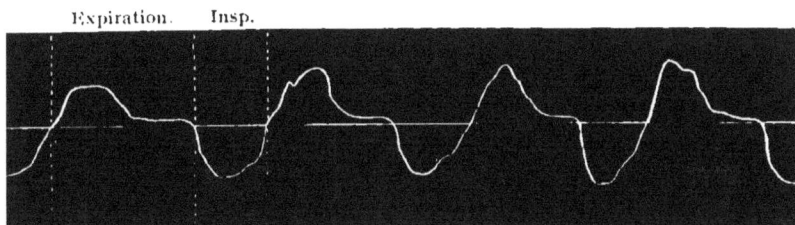

Expiration. Insp.

FIG. 2. Pneographic tracing — woman in corset.

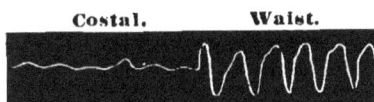

Costal. Waist.   Costal. Waist.

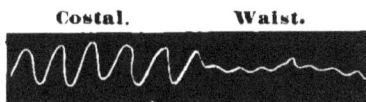

FIG. 3. Man.   FIG. 4. Woman in corset.

Costal   Waist.

FIG. 5. Chippewa Indian woman.

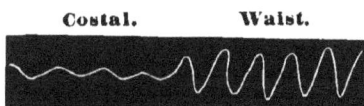

Costal. Waist.   Costal. Waist.

FIG. 6. Woman who never wore a corset.   FIG. 7. Man in corset.

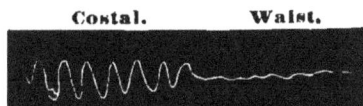

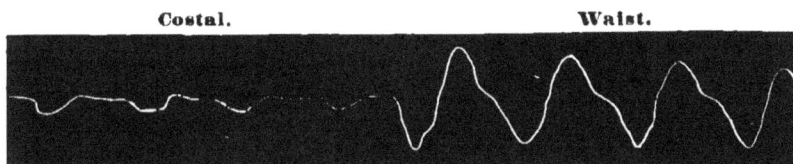

Costal.   Waist.

FIG. 8. Dog.

Costal.   Waist.

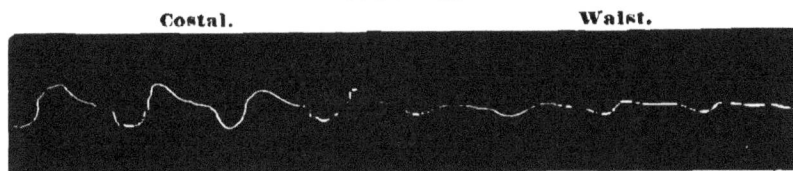

FIG. 9. Dog with corset on.

PLATE I.—BREATHING MOVEMENTS IN MAN AND DOG.

FIC. 5. Tracing obtained from a Chippewa Indian woman who had never worn a corset.

FIG. 6. Tracing furnished by a civilized woman who had never worn a corset. It will be noticed that the last two tracings have the same character as the tracing furnished by a healthy man, shown in Fig. 3.

FIG. 7. Tracing furnished by a man wearing a corset. This tracing is practically identical with the one shown in Fig. 4, furnished by a woman wearing a corset.

FIG. 8. Tracing obtained from a healthy dog.

FIG. 9. Tracing obtained from a dog with a corset on. By examination of the last two tracings it will be seen that a healthy dog breathes just as does a healthy man or healthy woman whose respiratory movements are unobstructed, and that a dog wearing a corset breathes as does a woman under the same circumstances, chiefly with the upper instead of the lower portion of the chest.

PLATE II.

FIGURE 1. Diagram of the trunk, showing position of the viscera, after Ziemssen. It will be noticed that the lower border of the stomach falls at a point about midway between the lower end of the sternum and the umbilicus.

FIG. 2. Diagram showing the action of the diaphragm, front view.

FIG. 3. Diagram showing the action of the diaphragm, side view.

PLATE III.

This plate is a representation of natural figures.

FIGURE 1. Outline of a healthy, well-developed man, thirty years of age.

FIG. 2. Outline of a well-developed woman, of twenty-six years.

FIG. 3. A Greek statue.

FIG. 4. Outline of an Italian model girl.

PLATE IV.

Outlines showing the effect of the corset in destroying the natural symmetry of the figure.

Figure 1. Side profile of a young woman who had once been addicted to tight lacing, but had greatly improved her figure by reforming her dress.

Fig. 2. Side profile of the same person with corset on. This tracing shows clearly the influence of the corset in destroying the natural dorsal curve, and producing protrusion of the lower abdomen.

Fig. 3. Front profile of a young woman of seventeen years who had never worn anything tight in her life. The two side sets of lines illustrate the form of the waist with a corset on and with it off. This young woman's waist measure was three inches less outside all her clothing with the corset on, than next the skin with the clothing removed.

Fig. 4. Side profile of the same person who furnished the preceding outline, showing the distortion of the figure and displacement of the internal viscera occasioned by tight lacing.

PLATE V.

Figure 1. The outline of a young woman who supposed she had always dressed healthfully, having worn a health-corset and suspended her clothing from her shoulders. The so-called health corset was tight and rigid with stays, and the skirt-bands were also tight and the skirts heavy. In consequence, the bowels and stomach were prolapsed, the lower border of the stomach reaching three inches below the umbilicus.

Fig. 2. The solid lines within the figure indicate the position of the stomach, liver, and right kidney. The dotted lines indicate the lower border of these organs when in normal position. The young woman was in most wretched health. She had suffered for many years from nervous dyspepsia, and also from pelvic congestion and displacement of the uterus and ovaries.

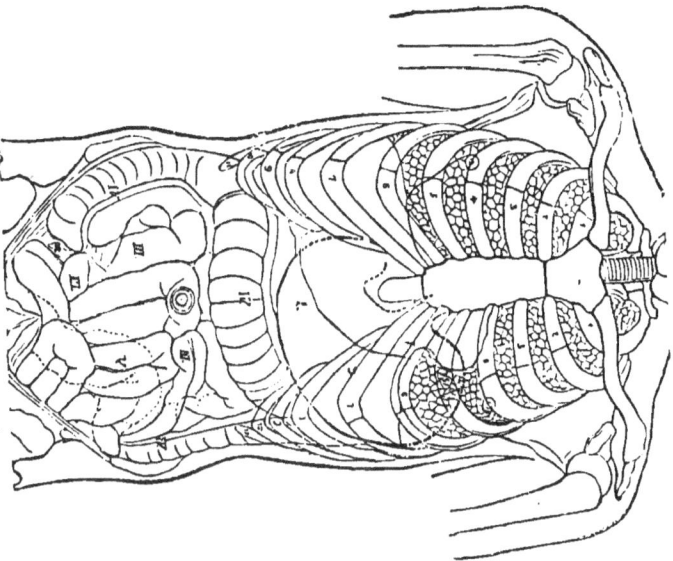

Fig. 1. (After Ziemssen.)

PLATE II.

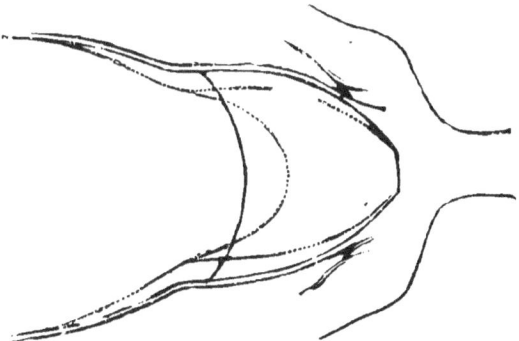

Fig. 2.

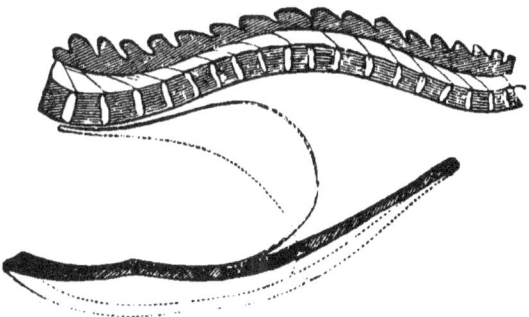

Fig. 3.

Reformed corset wearer.

Fig. 2. The same with corset.

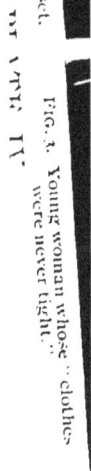

Fig. 3. Young woman whose " clothes were never tight."

Fig. 4. Side view of the same.

Fig. 2. Front view of the same.

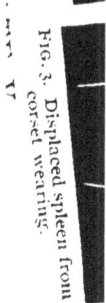

Fig. 3. Displaced spleen from corset wearing.

Fig. 4. Front view of the same.

FIG. 3. Outline of a woman of forty-two years, who, when a young woman, had compressed the waist for the purpose of getting rid of an enlarged spleen, which was finally crowded down below the waist-line, and, finding itself cut loose from its moorings, wandered about in all parts of the abdominal cavity. When first examined, the spleen — four or five times its normal size — lay between the uterus and the bladder, and was mistaken for a fibroid tumor. I discovered my error the next day, when I found the spleen lying several inches distant from its position of the day before.

FIG. 4. Front-view outline of the same patient. The solid lines S and L indicate the position in which the stomach and liver were found.

PLATE VI.

FIGURE I. Side-view profile of a woman who had suffered many years from nervous dyspepsia and a variety of nervous and pelvic troubles. She had consulted a great many physicians of eminence, but had found no relief.

FIG. 2. Front view of the same patient. The liver, stomach, and kidneys were several inches below their normal position. The displacement of the abdominal viscera was directly traceable to the influence of tight skirt-bands and heavy skirts. The patient was greatly improved in health by strengthening the abdominal walls, thus enabling them to hold the prolapsed organs more nearly in position.

FIG. 3. Side-view outline of a woman whose figure was greatly deformed, as the result of tight lacing and wearing heavy skirts suspended by tight bands. The liver occupied the position shown in the elliptical outline at L. The dotted line L indicates the normal position of the organ.

FIG. 4. This outline represents the side view of a man who had injured himself by wearing a belt. The liver was prolapsed two inches below the lowest ribs, and the stomach had fallen two or three inches lower. He was suffering from nervous dyspepsia, which was traced directly to the irritation set up in the prolapsed organs, as the result of their abnormal position.

PLATE VII.

FIGURE 1. Side profile of a woman of thirty-two years, who was suffering from uterine and ovarian prolapse, and prolapse of stomach, bowels, and kidneys. In consequence, she was a victim of nervous dyspepsia, general debility, and a variety of reflex nervous symptoms. Three months of careful regimen and training restored her to excellent health.

FIG. 2. Side profile of a man of forty years, a physician, in whom neglect of physical exercise and a bad position in standing and sitting, had caused prolapse of the stomach and bowels, and a numerous train of ills dependent upon these conditions.

FIG. 3. Side profile of a healthy, well-developed man. The same subject whose front profile is shown in Plate III.

FIG. 4. The solid outline represents the side profile of a man of forty-five years, who had suffered many years from nervous dyspepsia and a variety of nervous symptoms, which the treatment of a number of most excellent Eastern specialists had failed to relieve. A few weeks' physical training, in connection with other treatment, gave him the figure shown in the dotted outline, and restored him to such excellent health that he was able to return to his business enjoying better health than for many years. The principal causes of the deformity exhibited by this patient were wearing a belt when a young man, and neglect of physical exercise, his business habits having been sedentary from youth.

PLATE VIII.

FIGURE 1. Side-view outline of a woman of twenty-four years, who was suffering from nervous dyspepsia and prolapsus of uterus and ovaries.

FIG. 2. This outline represents the same young woman after a few months' physical training. She was restored to

Fig. 1. Effect of tight bands and heavy skirts.

Fig. 2. Front view of the same.

Fig. 3. Effects of corset-wearing and tight bands.

Fig. 4. Man who had worn a belt.

FIG. 1. Effects of corset and tight
lacing.

FIG. 2. The same person after
training.

PLATE VIII.

excellent health, and has since been able to engage regularly in the profession of nursing.

PLATE IX.

FIGURE 1. Side profile of a German peasant woman, twenty-nine years of age. Until twenty years of age, she was accustomed to carry heavy weights upon her head, often as much as ninety pounds, two or three miles without stopping to rest. She never had trained in gymnastics. It is a perfectly natural figure, and doubtless represents very nearly the ideal female form.

FIG. 2. Side profile of a woman of the same age, who, through neglect of muscular exercise, and by corset wearing and the wearing of tight bands and heavy skirts, had acquired the weak and deformed figure shown.

PLATE X.

FIGURE 1. A corset-choked woman (copied from a fashion plate).

FIG. 2. A healthfully dressed woman.

FIG. 3. This and the succeeding figure are intended to show the real origin of busts and bustles. The woman whose figure has been destroyed by corset-wearing, requires an artificial bust in front and a bustle behind to restore the natural curves of the figure.

FIG. 4. A woman with a natural figure who has no use for either bustles or artificial busts.

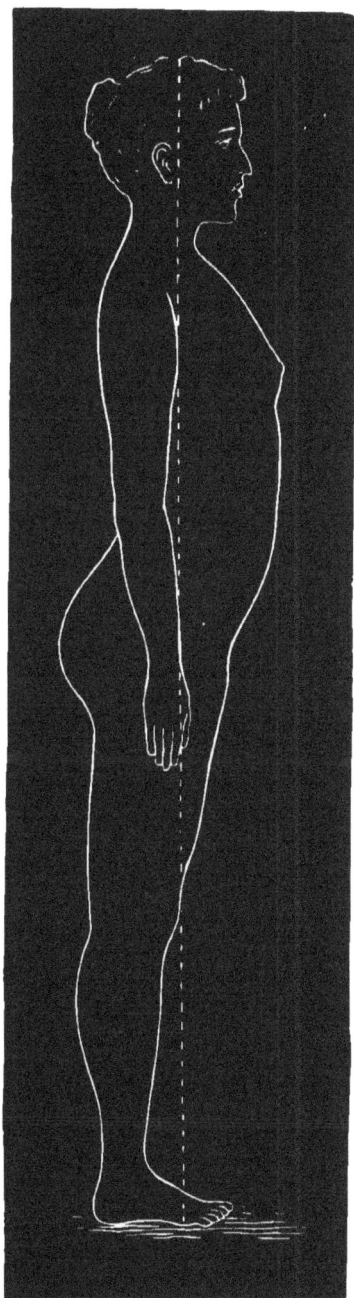

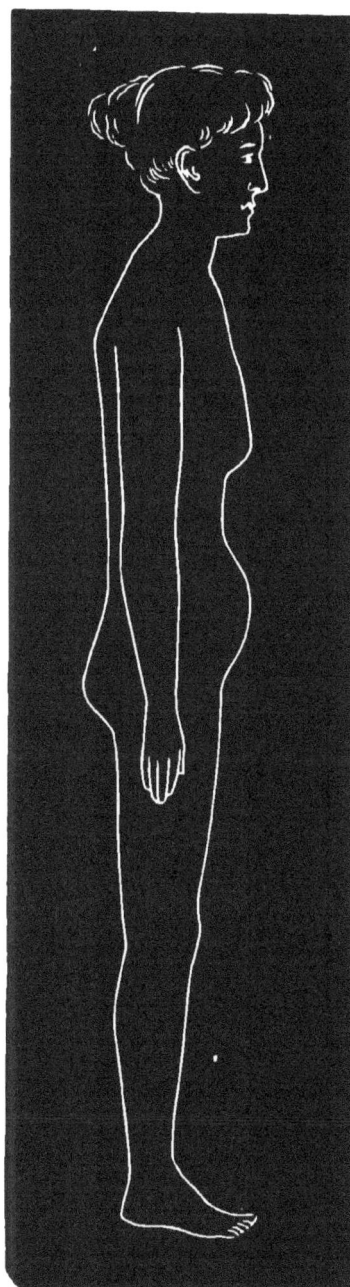

FIG. 1. A German peasant woman.　　FIG. 2. Effects of corset and tight bands, on an American woman of same age.

PLATE IX.

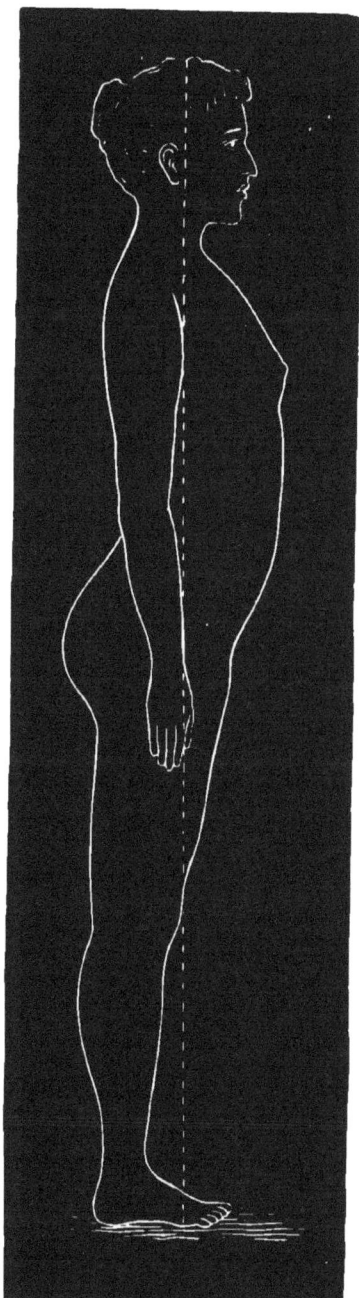

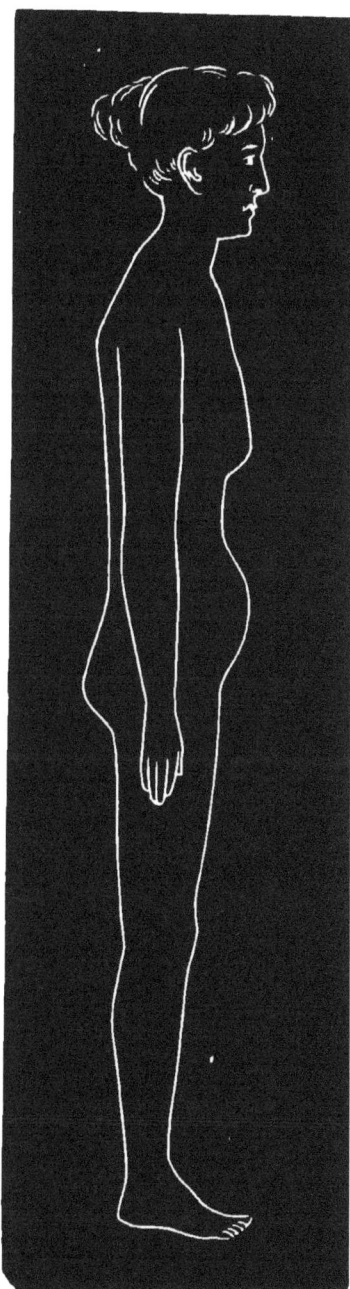

FIG. 1. A German peasant woman.   FIG. 2. Effects of corset and tight bands, on an American woman of same age.

PLATE IX.

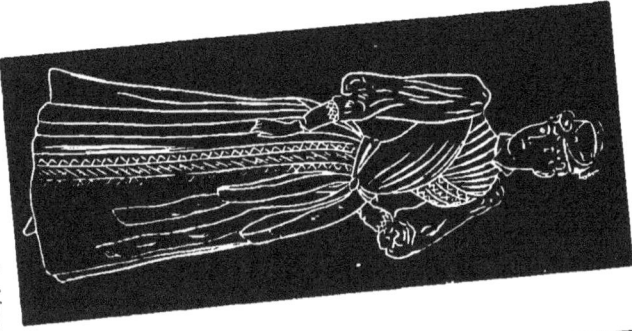

Fig. 1. Copied from a fashion plate.

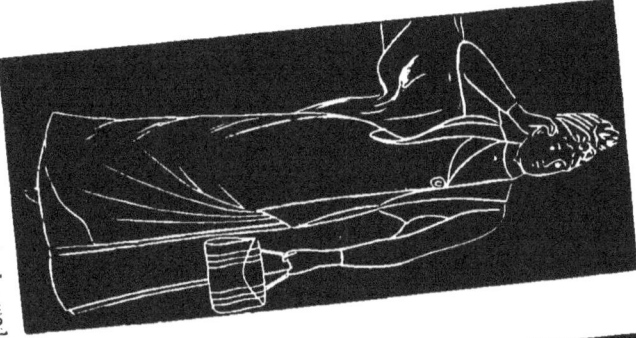

Fig. 2. A healthfully dressed woman.

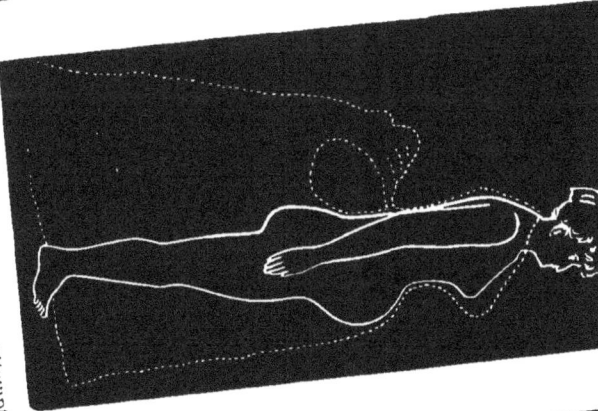

Fig. 3. An unnatural woman attempting to conceal defects.

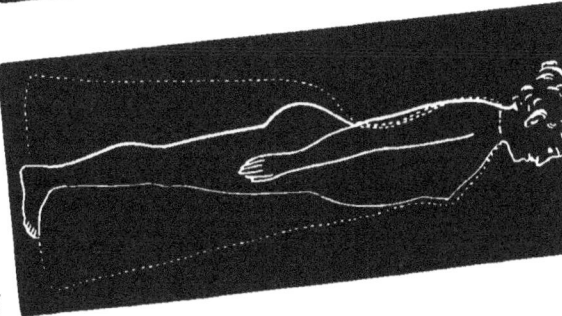

Fig. 4. A natural woman whose figure requires no appendages.

PLATE X.

www.ingramcontent.com/pod-product-compliance
Lightning Source LLC
Chambersburg PA
CBHW021638270326
41931CB00008B/1066